INTERNATIONAL INTERIOR DESIGN YEARBOOK 2016

2016 国际室内设计年鉴 ❻

OFFICE INSTITUTION

办公/社团

本书编委会 编

中国林业出版社
China Forestry Publishing House

图书在版编目（CIP）数据

国际室内设计年鉴.2016.办公、社团 /《国际室内设计年鉴》编委会编. -- 北京：中国林业出版社，2016.4

ISBN 978-7-5038-8454-2

Ⅰ.①国… Ⅱ.①国… Ⅲ.①办公建筑－室内装饰设计－世界－2016－年鉴 Ⅳ.①TU238-54

中国版本图书馆CIP数据核字(2016)第057495号

本书编委会

◎ 编委会成员名单
丛书主编：柳素荣
编写成员：陈向明　陈治强　董世雄　冯振勇　朱统菁
◎ 丛书策划：北京和易空间文化传播有限公司
◎ 特别鸣谢：《室内设计与装修》杂志社
◎ 装帧设计：北京睿宸弘文文化传播有限公司+LOMO红绿

中国林业出版社 · 建筑家居出版分社

责任编辑：王思源　纪亮

出版：中国林业出版社　（100009 北京西城区德内大街刘海胡同7号）
网址：lycb.forestry.gov.cn
电话：（010）8314 3518
发行：中国林业出版社
印刷：北京利丰雅高长城印刷有限公司
版次：2017年2月第1版
印次：2017年2月第1次
开本：230mm×305mm　1/16
印张：13.5
字数：200千字
定价：220.00元

CONTENTS 目录

004_ 导言
INTRODUCTION

007_ 办公
BUILDING

008_ 石家庄市科技服务中心
SHIJIAZHUANG TECHNOLOGY SERVICE CENTER
010_ 右脑建筑设计
RIGHT BRAIN ARCHITECTURAL DESIGN
014_ ANONIMO CORPORATE OFFICES
ANONIMO CORPORATE OFFICES
020_ KANTOOR DUPON办公室
KANTOOR DUPON OFFICE
024_ 卡昂大学医院中心
CAEN UNIVERSITY HOSPITAL CENTER
028_ STUDIO IPPOLITO FLEITZ GROUP
STUDIO IPPOLITO FLEITZ GROUP
032_ XMS 媒体艺廊·木石研、爻域工作实验室(工作空间)
KEEP INTERACTING XMS MEDIA
GALLERY & MOXIE+XXTPALAB (WORK SPACE)
038_ LEGACY BANK SCOTTSDALE AZ
LEGACY BANK SCOTTSDALE AZ
042_ 电子艺术中心
ARS ELECTRONICA CENTER
046_ 波特兰港总部
PORT OF PORTLAND HEADQUARTERS
050_ CALYON JAPAN
CALYON JAPAN
052_ 在库尔CATONAL银行格劳宾登的总办事处
THE HEAD OFFICE CATONAL BANK GRISONS IN CHUR
054_ PLAJER&FRANZ工作室办公室
OFFICE PLAJER&FRANZ STUDIO
060_ OFFICE VETRERIA AIROLDI
OFFICE VETRERIA AIROLDI
064_ 安德玛索卫浴总部大楼
ANDEMASUO BATH HEADQUARTER BUILDING
068_ ACBC OFFICE
ACBC OFFICE
076_ 黑弧奥美上海区域办公室
HEHUI AOMEI SHANGHAI OFFICE
082_ DOC BOY PUBLICATION
DOC BOY PUBLICATION
087_ WGV CUSTOMER SERVICE CENTRE
WGV CUSTOMER SERVICE CENTRE
090_ ERICSSON NANJING
ERICSSON NANJING
092_ 大都会大厦公共通道及大堂
METROPOLITAN TOWER PUBLIC PASSAGE & LOBBY
094_ 抽象水墨·解构
ABSTRACT INK·DECONSTRUCTION
098_ 上海鸿澜装饰设计工程有限公司
HONGHU DESIGN HONGLAN DECORATION
100_ 美丽华酒店工作区办公室
THE MIRA HONG KONG BACK OF HOUSE OFFICE
102_ 清祺书办公室
QINGQISHU OFFICE
106_ PLANECO CORPORATE OFFICES
PLANECO CORPORATE OFFICES
108_ CENTER FOR SUSTAINABILITY
CENTER FOR SUSTAINABILITY
110_ 鸿隆世纪广场
H·LCENTURY PLAZA
114_ 根植东方的非线性实性
ROOTED IN THE EAST NONLINEAR EXPERIMENTS
116_ 浙江深美装饰工程有限公司办公楼
ZHEJIAG SHENMEI DECORATION ENGINEERING CO.,LTD.OFFICE BUILDING
120_ 嘉丰涂料办公室
JIAFENG PAINT OFFICE
122_ 成都天府软件园服务中心办公楼
CHENGDU SOFTWARE PARK SERVICE CENTER OFFICE
126_ 深圳珂莱蒂尔服饰有限公司
SHENZHEN KORADIOR FASHION CO.,LTD
130_ 办公室
OFFICE

134_ 社团
INSTITUTION

135_ 航行
SAILING
138_ 台湾国家音乐厅
KHS ARTISTCENTER
142_ ICC KIDS PROGRAM 2010
ICC KIDS PROGRAM 2010
144_ 上海世博会土耳其国家馆
SHANGHAI EXPO TURKY HALL
154_ JERSEY BOYS THEATRE
JERSEY BOYS THEATRE
158_ HJØRRING CENTRAL LIBRARY
HJØRRING CENTRAL LIBRARY
162_ 上海世博会ALGERIAN国家馆
NIATIONAL ALGERIAN PAVILION AT WORLD EXPO SHANGHAI
164_ APPLEMORE COLLEGE PR CHOICE
APPLEMORE COLLEGE PR CHOICE
168_ DORNIER MUSEUM
DORNIER MUSEUM
172_ OBUSE LIBRARY MACHITOSHO TERRASOW
OBUSE LIBRARY MACHITOSHO TERRASOW
174_ ROC CARE
ROC CARE
176_ AN ABANDONED MOVIE PROP
AN ABANDONED MOVIE PROP
178_ 在第一次世界大战中的欧洲华工
CHINESE LABOURERS IN WORLD WARI,EUROPE
180_ THE WALT DISNEY FAMILY MUSEUM
THE WALT DIS NEY FAMILY MUSEUM
182_ KUA TRY-OUT LAB
KUA TRY-OUT LAB
184_ TOKYO BABY CAFE
TOKYO BABY CAFE
186_ 音乐和戏剧学院
CONSERVATOIRE DE MUSIQUE ET D'ART DRAMATIQUE
190_ 嘉禾青衣电影城
GOLDEN HARVEST TSING YI MULTIPLEX
192_ JAQUA
JAQUA
196_ DESIGN LAB
DESIGN LAB
198_ 韦斯特切斯特教室
WESTCHESTER REFORM TEMPLE&SCHOOL
200_ 嘉禾荃新天地电影城
GH CITYWALK CINEMA
202_ COWBOYS STADIUM
COWBOYS STADIUM
208_ 龙的DNA
DNA OF DRAGON

INTRODUCTION

/ RECORD THE EXCELLENCE PUBLISH THE QUINTESSNCE

/ 记录精英 传播经典

张先慧 /Zhang Xianhui

中国麦迪逊文化传播机构董事长
中国（广州、上海、北京）广告书店董事长
《麦迪逊丛书》主编
Chairman of China Madison Culture Communication Institutions
President of China(Guangzhou,Shanghai,Beijing) Advertising Bookshop
Chief Editor of "The Madison Series"

人的一生，绝大部分时间是在室内度过的。因此，人们设计创造的室内环境，必然会直接关系到人们室内生活、生产活动的质量，关系到人们的安全、健康、效率、舒适，等等。随着人们生活水平和审美能力的不断提高，人们更加注重生活环境的设计，对于室内设计的要求更加严格，需求也日益多样化、个性化。这就要求设计师一定要牢牢把握住时代的脉搏和潮流，以独特的眼光，运用与众不同的角度和表现手法进行创意性的设计，以满足人们对室内设计的需求。

然而，一件好的设计作品，不仅与设计师的专业素质和文化艺术素养等联系在一起，更离不开对他人成功经验的借鉴，为此，《国际室内设计年鉴2011》应运而生。

本年鉴秉持以中国大陆、中国香港、中国台湾为主，兼容其他国家与地区参与的原则，主张以创新与发展作为室内设计创作的主旋律，以科学与艺术相结合的审美眼光审视室内设计作品，力求打造全球最具影响力的室内设计行业年鉴，并使其成为各国设计师可以借鉴的经典书籍。

本年鉴征稿消息发出后，世界各地的设计机构与设计师都踊跃参与，大量投稿，投稿数量之多完全出乎我们的意料，最终本年鉴以一套五册的形式面世。

我们用年鉴的形式把当代最具价值的室内设计作品记录下来，传播开去，意在对室内设计文化予以保存的同时，也为读者提供了解当代设计状况及思想交流的平台。

"记录精英，传播经典"，这是《麦迪逊丛书》的宗旨。

希望业界朋友继续关注与支持我们。

One's lifetime mostly passes through in the interior. Therefore, the interior environment will directly involve quality of people's interior life, activities, people's safety, health, efficiency, comfort and so on. Along with the continuous improvement of people's living standard and aesthetic capacity, people pay more attention to living environment design, and their requirement for interior design is more strict, increasingly diverse and personalized. This requires that the designer firmly grasp the pulse of the times and trends, with the special insight, to use the different angles and methods of performance for creative design, in order to meet the needs of people's interior design requirement.

A masterpiece requires not only the link of the designer's professional quality and cultural art accomplishment, but also others' successful experiences. For this reason, "International Interior Design Yearbook 2010" is born at this right moment.

This yearbook gives priority to China Mainland, China Hong Kong and China Taiwan and pays much attention to other countries and areas, and it upholds the spirit that innovation and development should be the theme of interior design and that interior design works should be evaluated in a scientific and artistic perspective. Aiming at becoming the most influential global yearbook of interior design, this book is a classical one in the eyes of designers all over the world.

After the announcement of draft-collecting was spread, we have received so many contributions from the designers and organizations of almost every country. The number was so surprising. Finally, the yearbook is published in a set of five books.

We present the most valuable contemporary interior designs through publishing this yearbook in order to preserve the interior designing culture and provide a platform for readers to know about contemporary designing improvements and to communicate with each other.

"Record Excellence Works, Spread Classical Works" is the tenet of "Madison Series".

It will be our privilege to have your appreciation and support.

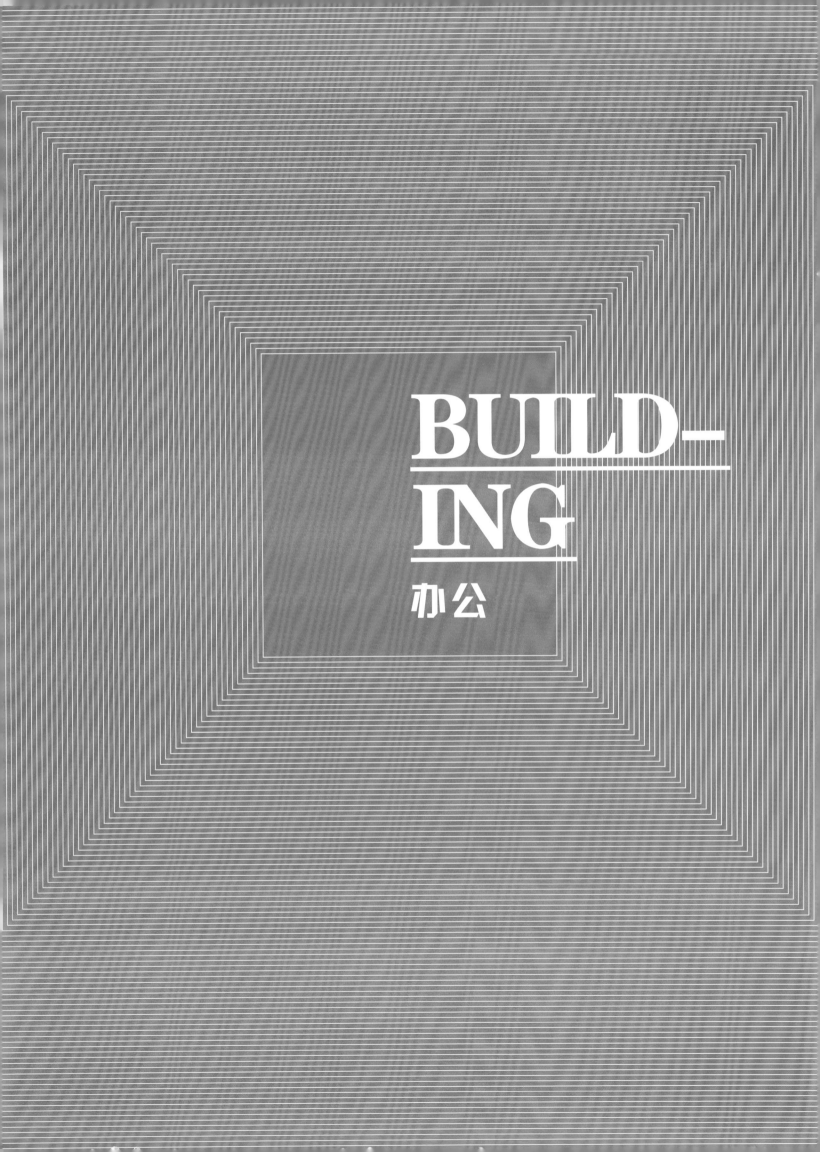

办公·BUILDING

石家庄市科技服务中心

SHIJIAZHUANG TECHNOLOGY SERVICE CENTER

项目资料:
设计单位:北京丽贝亚建筑装饰工程有限公司
设计总监(主创):甄健生
参与设计团队:郭瑞勇 林龙斌 陈晨
摄影师:徐盟
项目地址:石家庄市黄河大道136号
主要材料:花岗岩、大理石、木饰面、地毯、铝板

Project Information:
Design Unit: Beijing Libeiya Architectural Decoration Engineering Co., Ltd.
Design Director(Main Director): Zhen Jiansheng
Involved Design Team: Guo Ruiyong, Lin Longbin, Chen Chen
Photographer: Xu Meng
Project Address: No.136 Yellow River Avenue Shijiazhuang
Materials: Granite, marble, wood finishes, carpet, aluminum

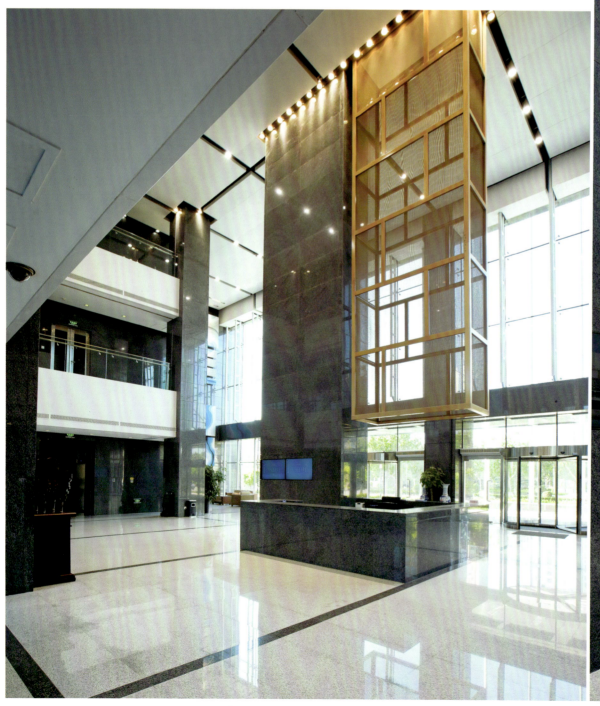

右脑建筑设计
RIGHT BRAIN ARCHITECTURAL DESIGN

项目资料：
设计单位：深圳市假日东方室内设计有限公司
项目地址：深圳市南山区高新科技园中区利苑路15号科兴生物谷2栋HHD假日东方国际设计师楼（整栋）
设计总监（主创）：洪忠轩
参与设计团队：王俊杰 马侠华
摄影师：陈中
主要材料：透光膜、墙纸、成晶镜框线、地毯、扪皮

Project Information：
Design Unit: Shenzhen holiday Oriental interior design Co., LTD
Project Address: Holiday Orient International Building 2 Kexing Bioy Valley Number 15 Liyuan Road New Technology Park Middle District Nanshan District Shenzhen
Design Director(Main Director): Hong Zhongxuan
Involved Design Team: Wang Junjie, Ma Xiahua
Photographer: Chen Zhong
Materials: Translucent membrane, wallpaper, into crystal frame line, carpet, palpable Paper

办公 · BUILDING

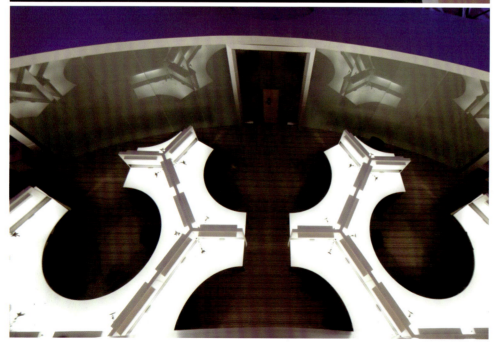

ANONIMO CORPORATE OFFICES

ANONIMO CORPORATE OFFICES

项目名称：
设计单位：DPGa
设计师：ARCH. DANIEL PEREZ GIL Director
摄影师：Hector Armando Herrera
客户：PLANECO
项目地址：Av. Paseo de la Reforma 1425, Lomas de Chapultepec, México, D.F
面积：540m²

Project Information:
Design Unit: DPGa
Designer: ARCH. DANIEL PEREZ GIL DIR.
Photographer: Hector Armando Herrera
Client: PLANECO
Address: Av. Paseo de la Reforma 1425, Lomas de Chapultepec, México, D.F.
Area: 540sqm

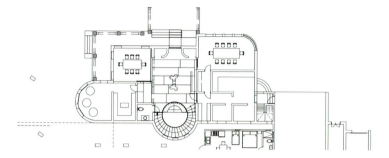

办公 · BUILDING

KANTOOR DUPON办公室
KANTOOR DUPONT OFFICE

项目资料：
设计单位：Studio Ramin Visch
项目团队：Ramin Visch, Femke Poppinga, Peter van der Geer(acoustic)
客户：Dupon Real Estate Development bv
简介：Office in a former Mayor Villa in Hoofddorp
面积：280m²
完成时间：2007年

Project Information:
Design Unit: Studio Ramin Visch
Projectteam: Ramin Visch, Femke Poppinga, Peter van der Geer(acoustic)
Client: Dupon Real Estate Development bv
Brief: Office in a former Mayor Villa in Hoofddorp
Area: 280sqm
Completion: 2007

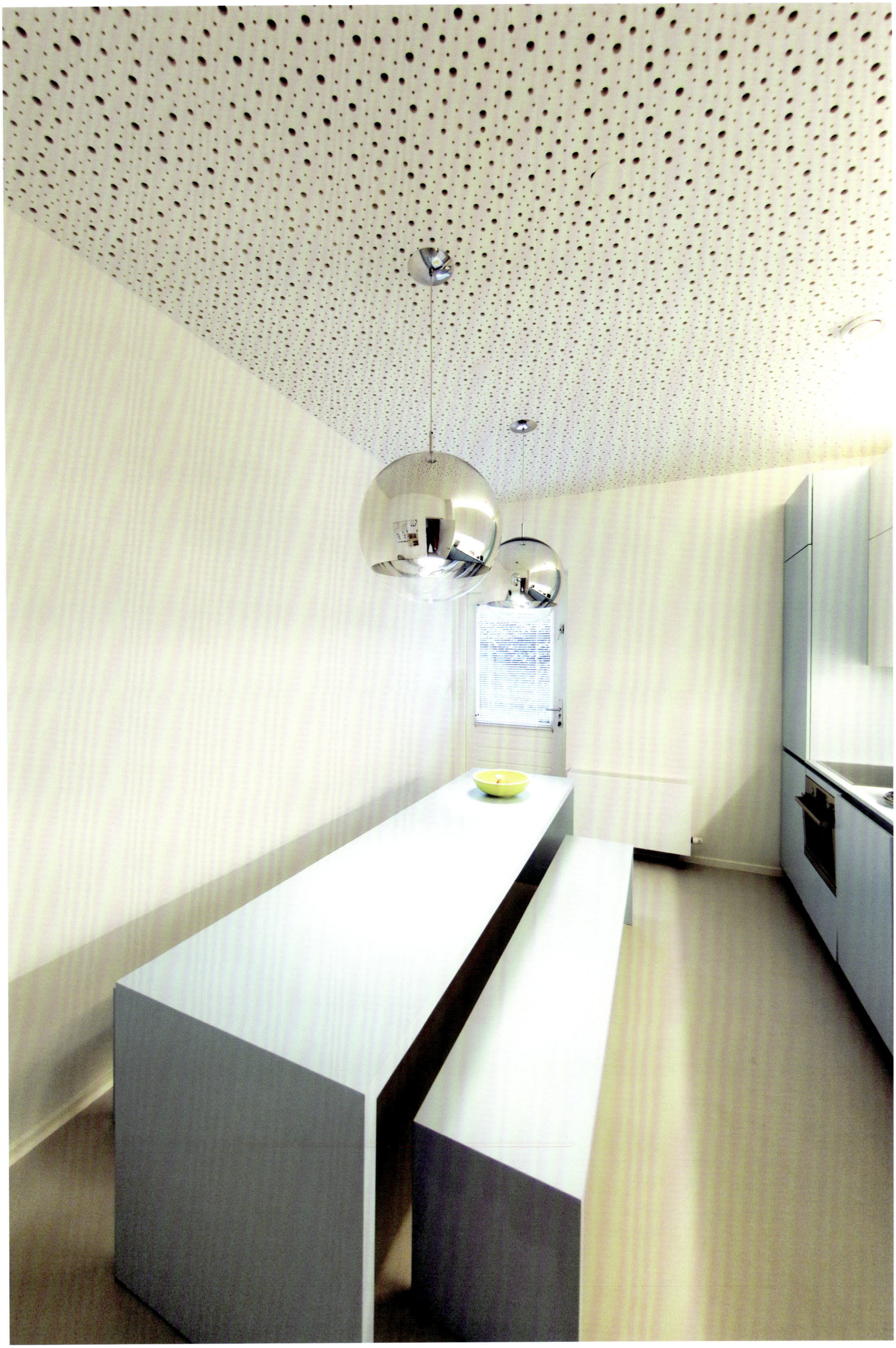

卡昂大学医院中心
CAEN UNIVERSITY HOSPITAL CENTER

项目资料：
设计单位：法国AS建筑工作室
设计师：ARCHITECTURE STUDIO

Project Information:
Design Unit: French AS Architecture Studio
Designer: ARCHITECTURE STUDIO

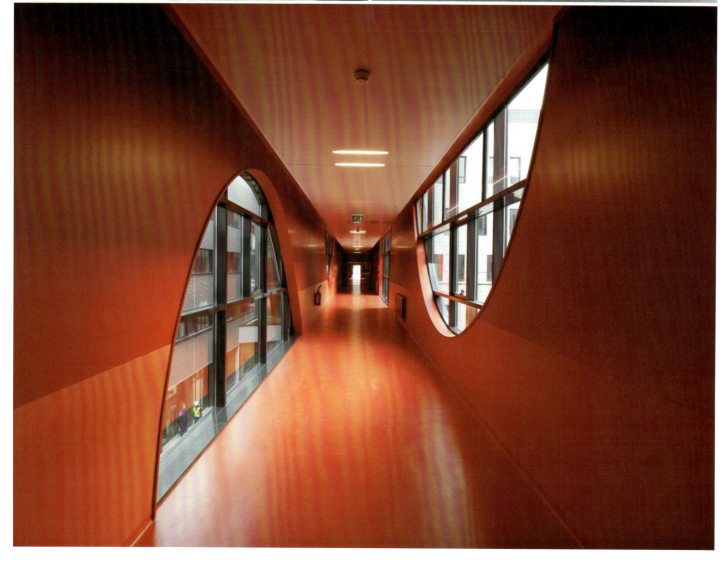

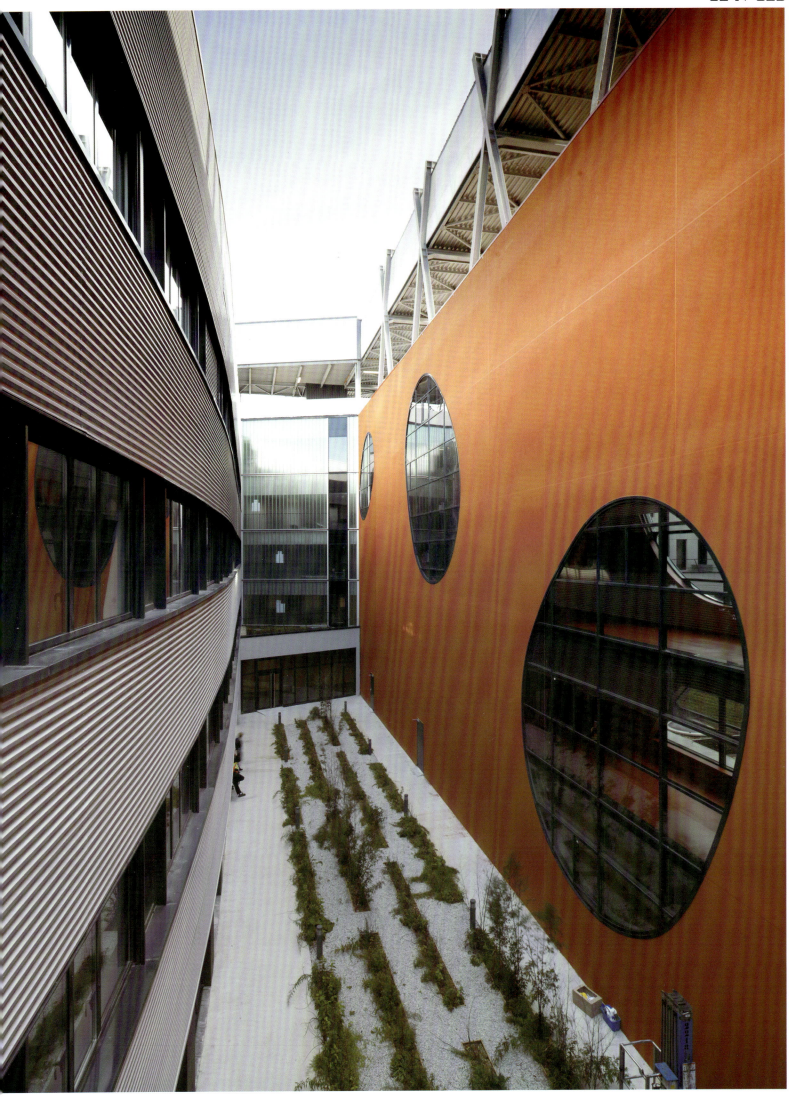

办公 · BUILDING

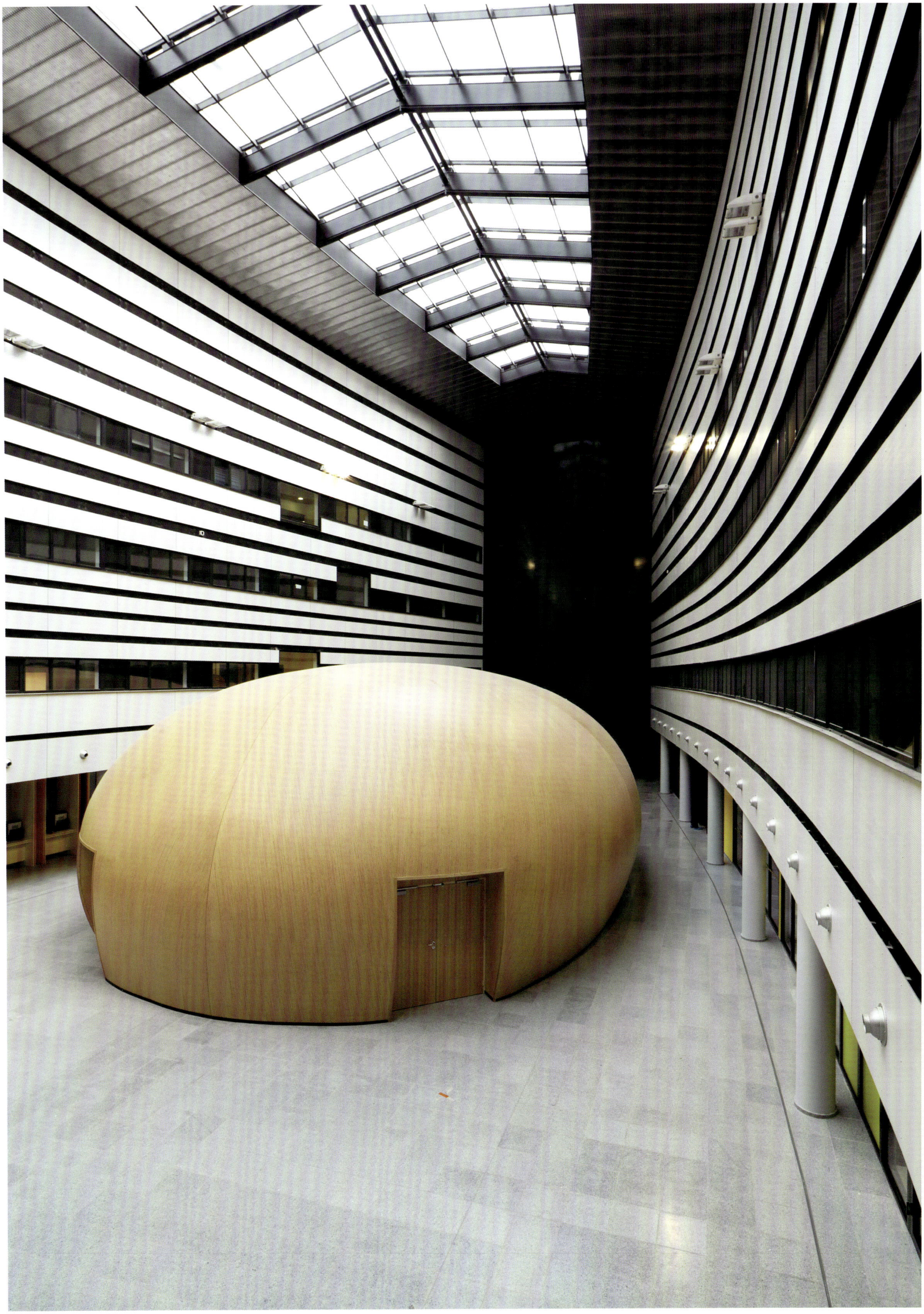

办公 · BUILDING

STUDIO IPPOLITO FLEITZ GROUP

STUDIO IPPOLITO FLEITZ GROUP

项目资料：
设计单位：Ippolito Fleitz Group GmbH Identity Architects
设计团队：Peter Ippolito, Gunter Fleitz, Sandra Böhringer, Inken Wellpottm, Julian Hensch, Christian Kirschenmann, Mathias Mödinger
摄影师：Zooey Braun
客户：Ippolito Fleitz Group GmbH
项目地址：Stuttgart, Germany
面积：480m²

Project Information：
Design Unit: Ippolito Fleitz Group GmbH Identity Architects
Design Team: Peter Ippolito, Gunter Fleitz, Sandra Böhringer, Inken Wellpottm, Julian Hensch, Christian Kirschenmann, Mathias Mödinger
Photographer: Zooey Braun
Client: Ippolito Fleitz Group GmbH
Project Address: Stuttgart, Germany
Area: 480sqm

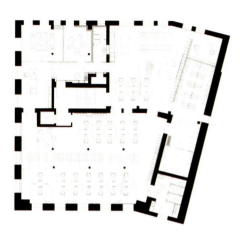

办公·BUILDING

XMS 媒体艺廊 木石研、爻域工作实验室(工作空间)

KEEP INTERACTING XMS MEDIA GALLERY & MOXIE + XXTRALAB(WORK SPACE)

项目资料：
设计单位：木石研室内建筑空间设计有限公司
设计总监（主创）：范赫铄
参与设计团队：木石研室内建筑空间设计有限公司
项目地址：台北市忠孝东路二段33号（10058）
主要材料：再生木、玻璃、钢刷木皮、互动多媒体、围篱软钢

Project Information：
Design Unit： MOXIE DESIGN
Design Director(Main Director)： Fan Heshuo
Involved Design Team： Mushiyan Interior Design Team
Materials： Recycled wood, glass, steel with brushed wood grain, interactive multimedia, soft fence net
Project Address： No.33 2 Duan Zhongxiao East Road Taipei
Materials： Wood, glass, Steel brush-wood, nteractive Multimedia, steel fence

办公·BUILDING

LEGACY BANK SCOTTSDALE AZ

LEGACY BANK SCOTTSDALE AZ

项目资料：
设计单位：Söhne & Partner Architects
建筑团队：Bartc, Prodinser, Traitpitsch
摄影师：Alexnder Koller, Severin Wirnis

Project Information:
Design Unit: Söhne & Partner Architects
Architect Team: Bartc, Prodinser, Traitpitsch
Photographer: Alexnder Koller, Severin Wirnis

办公 · BUILDING

电子艺术中心
ARS ELECTRONICA CENTER

项目资料：
设计单位：Treusch architecture ZT GmbH
创意总监：Andreas Treusch
建筑团队：N.Sailer, D.Kkkanovic, C.Jones, S.Scheffknechl, S.Ralzige, B.Scheffknechl, T.Schrittwieser, B.Harlung, C.Wellenzohn
摄影师：R.Steiner, A.Buchberger, A.Ehrenreich
面积：3336m²

Project Information:
Design Unit: Treusch architecture ZT GmbH
Creative Director: Andreas Treusch
Architect Team: N.Sailer, D.Kkkanovic, C.Jones, S.Scheffknechl, S.Ralziger, B.Scheffknechl, T.Schrittwieser, B.Harlung, C.Wellenzohn
Photographer: R.Steiner, A.Buchberger, A.Ehrenreich
Area: 3336sqm

办公 · BUILDING

办公 · BUILDING

波特兰港总部
PORT OF PORTLAND HEADQUARTERS

项目资料：
设计单位：ZGF Architects LLP

Project Information:
Design Unit: ZGF Architects LLP

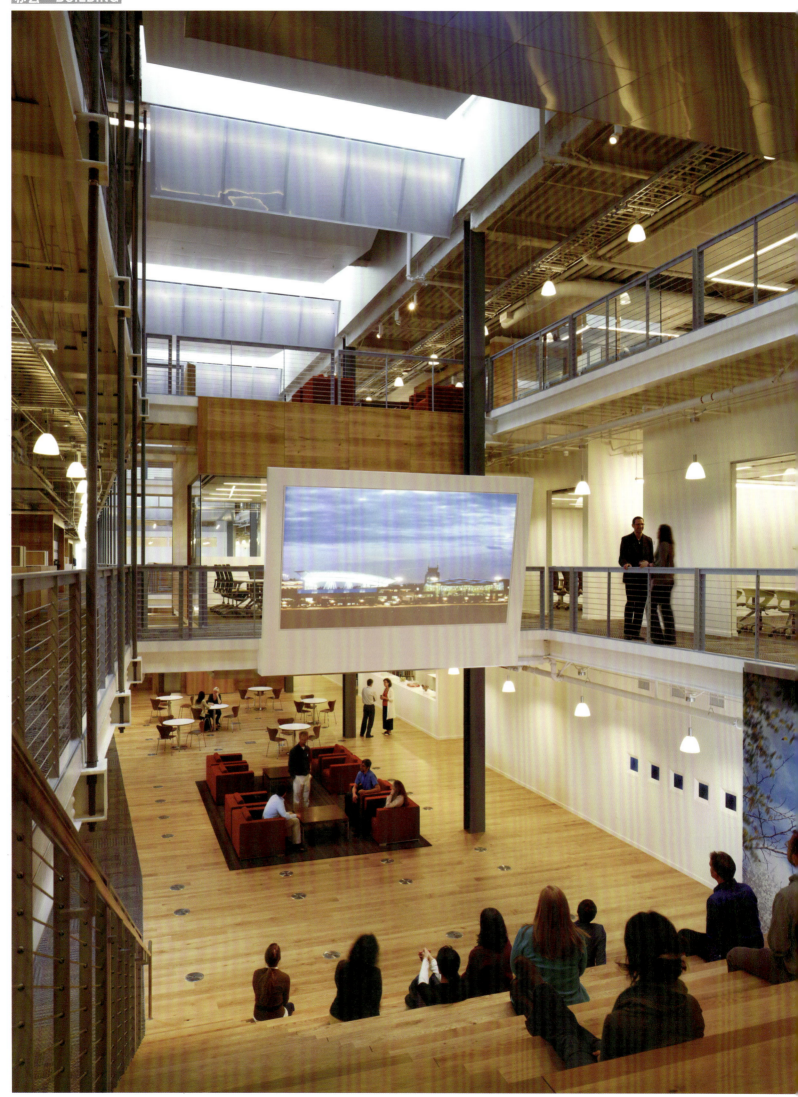

CALYON JAPAN
CALYON JAPAN

项目资料:
设计单位: Dasic Architects
建筑团队: George Dasic; Shunichi Tajima; Benigna Iwasaki, Reiko Fukushi and Tamaki Suzuki.
管理: John Lang LaSalle Koji Takenaka
助理建筑师: Ochs Architekten - Stefan Ochs
机电工程师: PT Morimura K. Ishiwatari, E. Maddison, D. Austin
资讯科技: PTS Jonathan Senycia
影音工程师: Eizo System
施工: Sumitomo R&D (Kajima Corp. 主承包商); Garde USP 室内承包商
Vectogramm: P+P Holzbau Gmbh - Thomas Poth
美术: Peter Cook. Tokyo
摄影师: Peter Cook. Tokyo
客户: Calyon Japan (Credit Agricole + Credit Lyonnais)
面积: 5000m²

Project Information:
Design Unit: Dasic Architects
Dasic Architects: George Dasic; Shunichi Tajima; Benigna Iwasaki, Reiko Fukushi and Tamaki Suzuki.
Proj.Manager: John Lang LaSalle Koji Takenaka
Associate Architect: Ochs Architekten - Stefan Ochs
M&E engineers: PT Morimura K. Ishiwatari, E. Maddison, D. Austin
IT: PTS Jonathan Senycia
AV engineers: Eizo System
Construction: Sumitomo R&D (Kajima Corp. Lead Contractor); Garde USP Interior Contractor
Vectogramm: P+P Holzbau Gmbh - Thomas Poth
Art: Peter Cook. Tokyo
Photography: Peter Cook. Tokyo
Client: Calyon Japan (Credit Agricole + Credit Lyonnais)
Area: 5000sqm

办公·BUILDING

在库尔CATONAL银行格劳宾登的总办事处
THE HEAD OFFICE CATONAL BANK GRISONS IN CHUR

项目资料：
设计单位：JüNGLING&HAGMANN ARCHIECTS
设计师：Dieter Juengling Andreas Hagmann
客户：GKB Chur

Project Information:
Design Unit: JüNGLING&HAGMANN ARCHIECTS
Designer: Dieter Juengling Andreas Hagmann
Client: GKB Chur

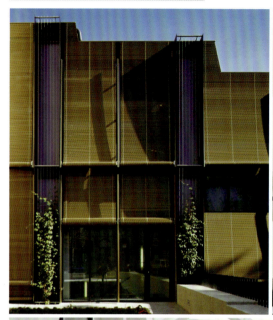

办公·BUILDING

PIAJER & FRAMZ工作室办公室
OFFICE PLAJER & FRANZ STUDIO

项目资料：
设计单位：plajer & franz studio
摄影师：ken schluchtmann
面积：1000m²

Project Information：
Design Unit: plajer & franz studio
Photographer: ken schluchtmann
Area: 1000sqm

OFFICE VETRERIA AIROLDI

OFFICE VETRERIA AIROLDI

项目资料：
设计单位：BURATTI + BATTISTON ARCHITECTS
设计师：Gabriele Buratti, Oscar Buratti, Ivano Battiston
建筑团队：Gabriele Buratti, Oscar Buratti, Ivano Battiston, Roberta Numi, Marco Vigano
摄影师：Marcello Mariana
项目地址：意大利San Giorgio su Legnano (Milano)

Project Information:
Design Unit: BURATTI + BATTISTON ARCHITECTS
Designer: Gabriele Buratti, Oscar Buratti, Ivano Battiston
Architect Team: Gabriele Buratti, Oscar Buratti, Ivano Battiston, Roberta Numi, Marco Vigano
Photographer: Marcello Mariana
Poject Address: San Giorgio su Legnano (Milano), ITALY

安德玛索卫浴总部大楼
ANDEMASUO BATH HEADQUARTER BUILDING

项目资料：
设计单位：墨象设计顾问有限公司
设计师：梁宇曦
项目地址：佛山市季华西路中国陶瓷城
建筑面积：629m²
主要材料：清玻璃、木饰面、亮光漆

Project Information：
Design Unit: Moxiang Design Consulting Co., Ltd.
Designer: Liang Yuxi
Project Address: Foshan Jihua West Road China china town
Area: 629sqm
Materials: Clear glass, Wood finishes, Varnish

ACBC OFFICE

项目资料：
设计单位：Pascal Arquitectos
摄影师：Jaime Navarro
项目地址：Torre Arcos Bosques, Paseo de los Tamarindos n°400 A, Colonia. Bosques de las Lomas, Delegación Cuajimalpa en la Ciudad de México
面积：512m²
完成时间：2010年

Project Information:
Design Unit: Pascal Arquitectos
Photographer: Jaime Navarro
Project Address: Torre Arcos Bosques, Paseo de los Tamarindos n°400 A, Colonia. Bosques de las Lomas, Delegación Cuajimalpa en la Ciudad de México
Area: 512sqm
Compeletion: 2010

黑弧奥美 上海区域办公室
HEHUI AOMEI SHANGHAI OFFICE

项目资料：
设计单位：上海塞赫建筑咨询
设计师：王士鲥
摄影师：Kim Yee
项目地址：上海市 局门路 八号桥创意园二期
面积：约660m²
主要材料：高密度板+亚光清漆完成面（楼梯+橱柜+家具）、灰色PVC地面卷材、白色环氧自流平地坪微晶石台面、乳白色防火板桌面、千层板桌板、明镜墙面处理、钢化玻璃隔断、高密度板现场喷漆门套+窗套+踢脚线、10x10mm铝合金U型条细部收边、乳胶漆墙面
完成时间：2010年2月

Project Information:
Design Unit: Shanghai Saihe Architectural Consultants
Designer: Wang Shihe
Photographer: Kim Yee
Project Address: Shanghai Jumen Road No. 8 Bridge Creative Park Stage Two
Area: 660sqm
Major Materials: MDF with Non-Gloss Clear Epoxy Finish, MDF with Non-Gloss Whit Paint Finish, White Epoxy Floor, PVC Carpet Floor Surface, Plywood Table Top with Formica Surface Layer, 5mm Mirror Panels, MDF Doors + Frames + Skirting, 12mm Tempered Clear Glass Partitions、8mm Cork Panels、Wall + Ceiling Paint Finishes
Completion: Feb, 2010

5th Floor Floor Plan

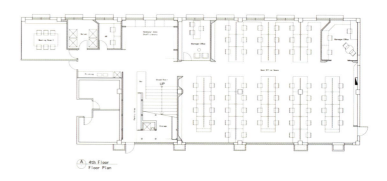
4th Floor Floor Plan

BlackArc Ogilvy
黑弧奥美房地产整合行销传播集团
广告 · 公关 · 互动 · 渠道 · 环境 · 数据库
BlackArc Ogilvy

DOC BOY PUBLICATION
DOC BOY PUBLICATION

项目资料:
设计单位: Studio Ramin Visch

Project Information:
Design Unit: Studio Ramin Visch

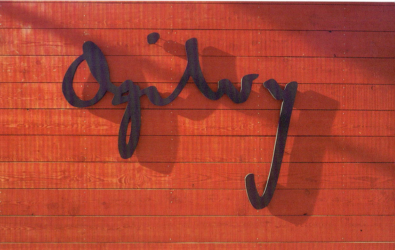

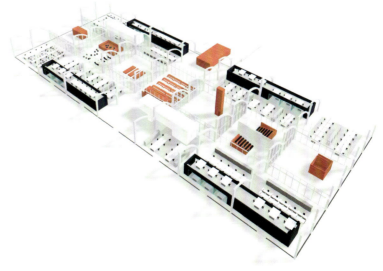

WGV CUSTOMER SERVICE CENTRE

WGV CUSTOMER SERVICE CENTRE

项目资料：
设计单位： Ippolito Fleitz Group GmbH Identity Architects
设计团队： Peter Ippolito, Gunter Fleitz, Fabian Greiner, Christian Kirschenmann, Vincent Gabriel
摄影师： Zooey Braun
客户： Wurttembergische Gemeinde.
项目地址： FeinstraBe 1, 70178 Stuttgart
面积： 1000m²

Project Information:
Design Unit: Ippolito Fleitz Group GmbH Identity Architects
Design Team: Peter Ippolito, Gunter Fleitz, Fabian Greiner, Christian Kirschenmann, Vincent Gabriel
Photographer: Zooey Braun
Client: Wurttembergische Gemeinde.
Projet Address: FeinstraBe 1, 70178 Stuttgart
Area: 1000sqm

ERICSSON NANJING

ERICSSON NANJING

项目资料：
设计单位：DPWT设计有限公司
设计师：Arthur Chan
摄影师：Mr. Diamond Chan
项目地址：南京江宁区胜利路89号
面积：2800m²
完成时间：2010年3月

Project Information:
Design Unit: DPWT Design Ltd
Designer: Arthur Chan
Photographer: Mr. Diamond Chan
Project Address: No.89, Shengli Road, Jiangning District, Nanjing
Area: 2800sqm
Completion: March, 2010

大都会大厦公共通道及大堂
METROPOLITAN TOWER PUBLIC PASSAGE & LOBBY

项目资料：
设计单位：ROGERS MARVEL ARCHITECTS, PLLC
建筑团队：Rogers Marvel
摄影师：©Paul Warchol
项目地址：纽约

Project Information:
Design Unit: ROGERS MARVEL ARCHITECTS, PLLC
Architecture Team: Rogers Marvel
Photographer: ©Paul Warchol
Project Address: New York, NY

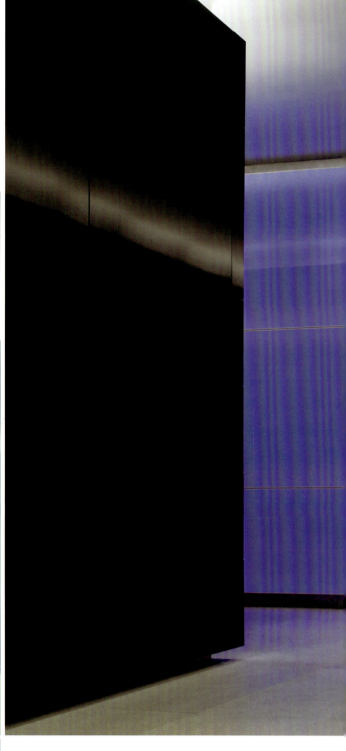

抽象水墨·解构
ABSTRACT INK · DECONSTRUCTION

项目资料：
设计单位：鸿扬集团陈志斌设计事务所
主设计师：陈志斌
参与设计团队：李智勇 谭丽 彭辉
摄影师：周文 管盼星
项目地址：长沙市袁家岭鸿扬大厦
面积：118m²
主要材料：加州金麻火烧板石材、不锈钢帘、雕花灰镜、彩绘、LG地胶
完成时间：2009年6月

Project Information:
Design Unit: Hongyang Group Chenzhibin Design Office
Main Designer: Chen Zhibin
Involed Design Team: Li Zhiyong, Tan Li, Peng Hui
Photographer: Zhou Wen, Guan Panxing
Project Address: Hi-run Building, Yujialing, Changsha, China
Area: 118sqm
Meterials: Giallo California baked board stone, stainless steel curtains, gray mirror with carved patterns, coloured drawing, LG floor gum
Completion: June 2009

上海鸿澜装饰设计工程有限公司
HONGHU DESIGN HONGLAN DECORATION

项目资料：
设计单位：上海鸿澜装饰设计工程有限公司
设计总监：黄宇 朱娟娟
设计团队：鸿鹄设计
摄影师：朱娟娟
项目地址：常州市嘉宏国际大厦
主要材料：三辉硅藻泥、白色帷幔、玻璃、星空顶

Project Information：
Design Unit: Shanghai Honglan Decoration Design Engineering Co., Ltd
Design Director: Huang Yu, Zhu Juanjuan
Design Team: Honghu Design
Photographer: Zhu Juanjuan
Project Address: Changzhou Jiahong International Mansion
Materials: Sanhui diatom mud, white curtains, glass, Stars top

办公·BUILDING

美丽华酒店工作区办公室

THE MIRA HONG KONG BACK OF HOUSE OFFICE

项目资料：
设计单位：DPWT Design Ltd.
设计团队：陈轩明 郑雅文 阮洁 钟玉枝
客户：美丽华集团
项目地址：香港九龙尖沙咀弥敦道
面积：790m²
设计风格：现代、有活力、启发灵感
主要材料：碳色地毯、铝、不锈钢、重点强调的暖色调子玻璃墙与塑料薄膜照明
完成时间：2009年4月

Project Information:
Design Unit: DPWT Design Ltd.
Designer: Arthur Chan, Summer Zheng, Emma Ruan and Gigi Chung
Client: The Miramar Group
Project Address: 118-130 Nathan Road, Tsim Sha Tsui, Hong Kong
Area: 790sqm
Design Style: Modern, Energetic, and Inspiring
Materials: Charcoal carpet, aluminium and stainless steel and vibrant,
warm tone colours' spray paint glass wall with featured lighting for accent highlight.
Completion: April 2009

项目资料：
设计单位：DPWT Design Ltd.
设计团队：陈轩明 郑雅文 阮洁 钟玉枝
客户：美丽华集团
项目地址：香港九龙尖沙咀弥敦道
面积：790m²
设计风格：现代、有活力、启发灵感
主要材料：碳色地毯、铝、不锈钢、重点强调的暖色调子玻璃墙与塑料薄膜照明
完成时间：2009年4月

办公·BUILDING

清祺书办公室
QINGQISHU OFFICE

项目资料：
设计单位：深圳华空间机构
设计师：熊华阳
项目地址：福田保税区长平商务大厦
面积：1200m²
工程造价：50万
主要材料：白铁皮、原木、水泥、微晶石

Project Information：
Design Unit: Shenzhen Hua Space Constitution
Designer: Xiong Huayang
Project Address: Futian Baoshui District Changping Business Building
Area: 1200sqm
Cost: 50Millions
Materials: Tin, timber, cement, ceramic stone

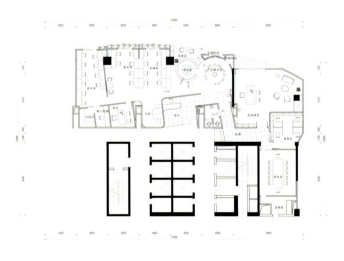

PLANECO CORPORATE OFFICES

PLANECO CORPORATE OFFICES

项目资料：
设计单位：DPGa
设计师：ARCH. DANIEL PEREZ GIL DIR.
摄影师：Hector Armando Herrera
客户：PLANECO
项目地址：Periferico sur #2349 PH, Col. Atlamaya, Delegación Alvaro Obregon, México, D.F.
面积：540m²

Project Information：
Design Unit: DPGa
Designer: ARCH. DANIEL PEREZ GIL DIR.
Photographer: Hector Armando Herrera
Client: PLANECO
Project Address: Periferico sur #2349 PH, Col. Atlamaya, Delegación Alvaro Obregon, México, D.F.
Area: 540sqm

CENTER FOR SUSTAINABILITY

CENTER FOR SUSTAINABILITY

项目资料：
设计单位：Totems Communication
创意总监：Peter van Lier
建筑团队：Simone Königshausen
摄影师：Totems Communication
项目地址：荷兰埃茵霍温

Project Information:
Design Unit: Totems Communication
Creative Director: Peter van Lier
Architect Team: Simone Königshausen
Photographer: Totems Communication
Project Address: Eindhoven, the Netherlands

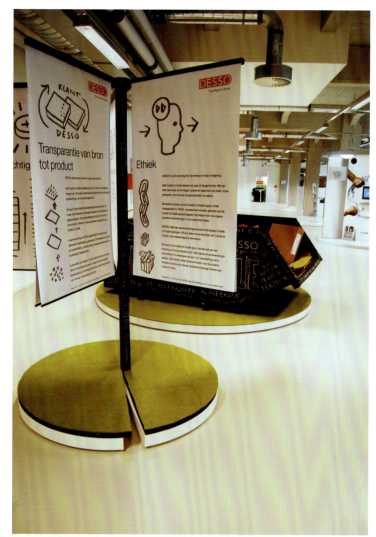

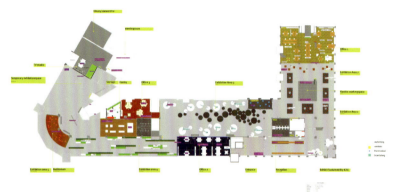

项目资料：
设计单位：Totems Communication
创意总监：Peter van Lier
建筑团队：Simone Königshausen
摄影师：Totems Communication
项目地址：荷兰埃茵霍温

鸿隆世纪广场
H·L CENTURY PLAZA

项目资料:
设计单位：深圳创域设计有限公司
项目地址：深圳市罗湖区和平路
建筑面积：503m²
主要材料：白洞石、超白玻璃、金影木、黑镜钢、地毯
完成时间：2009年11月

Project Information:
Design Unit: Shenzhen Chuangyu Design Co., Ltd.
Project Address: Shenzhen Luohu District Peace Road
Area: 503sqm
Material: white hole stone, super-white glass, gold shadow wood, black mirror steel, Rugs
Completion: November, 2009

雕塑小景

根植东方的非线性实验

ROOTED IN THE EAST NONLINEAR EXPERIMENTS

项目资料：
设计单位：鸿扬集团陈志斌设计事务所
主设计师：陈志斌
参与设计团队：李勇 谭丽 彭辉
摄影师：周文 管盼星
项目地址：长沙市袁家岭鸿扬大楼
面积：28m²
主要材料：灰色墙漆、银色油漆、LG地胶
完成时间：2009年6月

Project Information：
Design Unit: Hongyang Group Chenzhibin Design Office
Main Designer: Chen Zhibin
Involed Design Team: Li Yong, Tan Li, Peng Hui
Photographer: Zhou Wen, Guan Panxing
Project Address: Hi-run Building, Yujialing, Changsha, China
Area: 28sqm
Meterials: Gray wall paint, silver paint, LG floor gum
Completion: June 2009

浙江深美装饰工程有限公司办公楼

ZHEJIANG SHENMEI DECORATION ENGINEERING CO., LTD. OFFICE BUILDING

项目资料：
设计单位：杭州易和室内设计有限公司
设计总监（主创）：马辉
参与设计团队：莫雅萍 卞建辉
项目地址：宁波慈城
主要材料：地毯、红橡、大理石

Project Information:
Design Unit: Hangzhou Yihe Interior Design Co., Ltd.
Design Director(Main Director): Ma Hui
Involved Design Team: Mo Yaping, Bian Jianhui
Project Address: Ningbo Cicheng
Materials: Carpet, Red Oak, Marbles

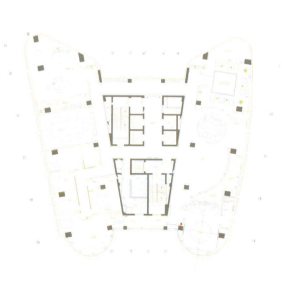

嘉丰涂料办公室
JIAFENG PAINT OFFICE

项目资料：
设计单位：福州柯恩罗伊景观设计有限公司
设计总监（主创）：滕家锋
参与设计团队：KR一组
摄影师：李玲玉
项目地址：福州
主要材料：米黄金刚板、青石板、有色乳胶漆

Project Information：
Design Unit: Fuzhou Roy Cohen Landscape Design Co., Ltd.
Design Director(Main Director): Teng Jiafeng
Involed Design Team: Group RK
Photographer: Li Lingyu
Project Address: Fuzhou
Materials: Rice-Yellow Jingang Plate, Blue Stone Plate, Colored Latex Paint

平面布置图

成都天府软件园服务中心办公楼

CHENGDU SOFTWARE PARK SERVICE CENTER OFFICE

项目资料：

设计单位：北京银坐标室内装饰设计有限公司
设计总监：刘立伟
参与设计团队：孙恺 董维倩 邱爱成 田钢 潘岫峰
项目地址：成都市高新技术开发区
主要材料：水泥板、胡桃木饰面、自流平等

Project Information：

Design Unit: Beijing Silver Zuobiao Interior Decoration Design Co.,Ltd.
Design Director: Liu Liwei
Involed Design Team: Sun Kai, Dong Weiqian, Qiu Aicheng, Tian Gang, Pan Xiufeng
Project Address: Chengdu Gaoxin Technology Development Zone
Materials: Cement board, walnut finish, self-leveling, etc.

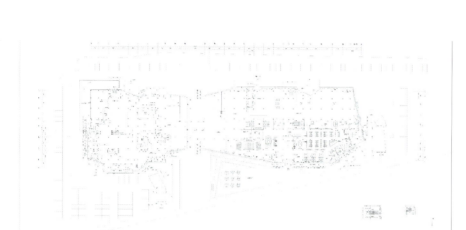

办公·BUILDING

深圳珂莱蒂尔服饰有限公司

SHENZHEN KORADIOR FASHION CO.,LTD.

项目资料：
设计单位：深圳瑞和建筑装饰设计研究院
设计总监（主创）：陈任远
参与设计团队：陈任远 侯江涛 吴明
摄影师：陈任远
项目地址：深圳市福田区车公庙泰然九路红松大厦B座七楼
主要材料：石材（黄洞石、灰木纹石、黑木纹石等）、地砖（金属转、地砖等）、饰面板（白象木、条纹吸音板等）、涂料（乳胶漆等）、玻璃（灰色钢化玻璃、夹丝玻璃镀膜玫瑰图案等）、镜（银镜、灰镜等）、金属（不锈钢）、地毯（方块毯等）、木板（橡木实木复合木地板、防腐木地板等）、墙纸、皮革、贝壳马赛克

Project Information：
Design Unit: Shenzhen Ruihe Architectural Decoration Design Institute
Design Director(Main Director): Chen Renyuan
Involed Deisgn Team: Chen Renyuan, Hou Jiangtao, Wu Ming
Photographer: Chen Renyuan
Project Address: Floor Seven Building B, Hongsong Building No.9 Tairan Nine Road, Chegong Temple Futian District Shenzhen
Materials: Stone (Yellow-cave Stone, Gray Wood Stone, Black Wood Stone etc.), Tiles (Metal Tiles, Tiles etc.), Veneer (White Elephant Wood, Acoustic Panel Stripes etc.), Painting (Latex Paint etc.), Glass (Grey Tempered Glass, Wired glass coating rose pattern etc.), Mirror (Silver Mirror, Grey Mirror etc.), Metal (Stainless Steel), Carpet (Blanket etc.), Wood (Oak Solid Wood Flooring, Antiseptic Wooden Flooring etc.), Wallpaper, Leather, Shell Mosaic

辨公室
OFFICE

办公·BUILDING

项目资料：
设计单位：杨焕生建筑室内设计事务所
设计总监：杨焕生
参与设计团队：王慧静 郭士豪
摄影师：刘俊杰
项目地址：台中大里
主要材料：半抛石英砖、全木皮染灰、喷漆白、窗花门片、灰大理石

Project Information：
Design Unit: Yang Huansheng Architectural Interior Design Office
Design Director: Yang Huansheng
Involed Design Team: Wang Huijing, Guo Shihao
Photographer: Liu Junjie
Project Address: Taizhong Dali
Materials: Half-cast porcelain tile, full veneer dyed gray, white paint, window bars and grilles, gray marble

航行
SAILING

项目资料:
设计单位: 芮马设计工作室

Project Information:
Design Unit: Ruima Design Studio

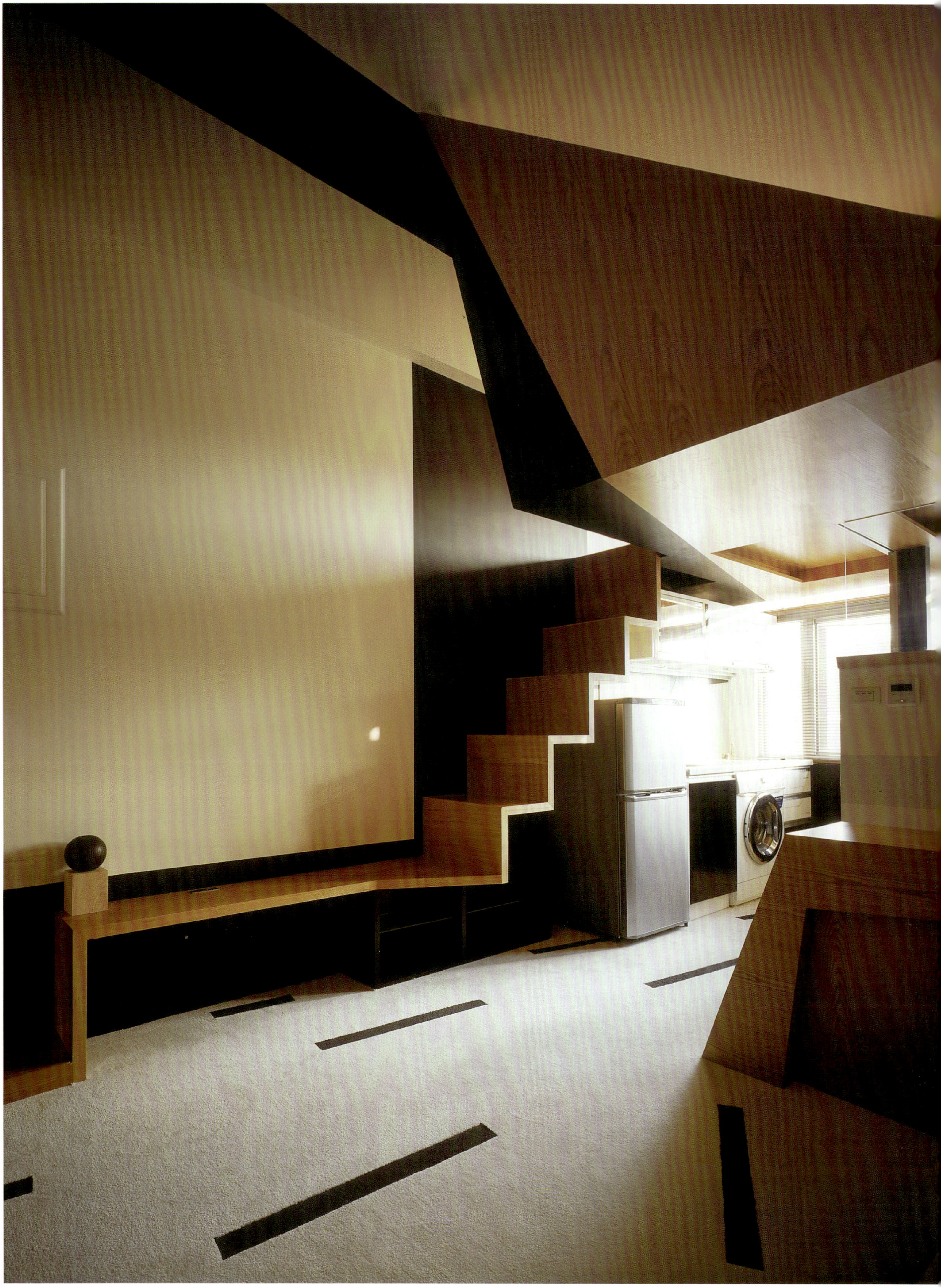

台湾国家音乐厅
KHS ARTIST CENTER

项目资料：
设计单位：芮马设计工作室

Project Information:
Design Unit: Ruima Design Studio

PLAN

ICC KIDS PROGRAM 2010
ICC KIDS PROGRAM 2010

项目资料：
设计单位：SOUP DESIGN
设计师：Nobuaki Doi
参与者：Izumi OKayasu(Lighting Design),Shingo Ishikawa, Masato Shimakawa
摄影师：Takumi Ota
项目地址：日本163-1404东京,新宿区Nishishinjuku3 - 20 - 2东京歌剧城大厦4楼
主要材料：非织造布
完成时间：2010年8月

Project Information：
Design Unit: SOUP DESIGN
Designer: Nobuaki Doi
Participants: Izumi OKayasu(Lighting Design),Shingo Ishikawa, Masato Shimakawa
Photographer: Takumi Ota
Project Address: Tokyo Opera City Tower 4F, 3-20-2 Nishishinjuku, Shinjuku-ku, Tokyo 163-1404 Japan
Materials: Nonwoven fabric
Completion: August, 2010

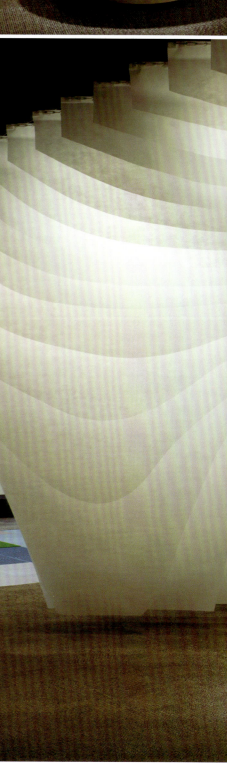

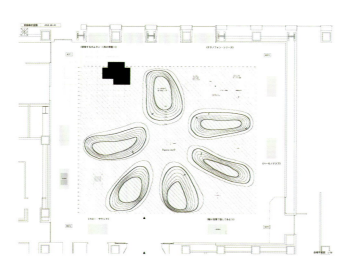

社团·INSTITUTION

上海世博会土耳其国家馆

SHANGHAI EXPO TURKY HALL

项目资料：
设计单位：Z+G INTERNATIONAL
设计总监：Ignacio Goded 钟德跃
参与设计团队：Ateam 国际建筑设计事务所（加拿大）
摄影师：吴牧
项目地址：上海世博会C片区
主要材料：SPUA塑性材料

Project Information：
Design Unit: Z+G INTERNATIONAL
Design Director: Ignacio Goded, Zhong Deyue
Involed Design Team: Ateam International Architectural Design Office (Canada)
Photographer: Wu Mu
Project Address: Shanghai Expo Area C
Materials: SPUA Plastic Material

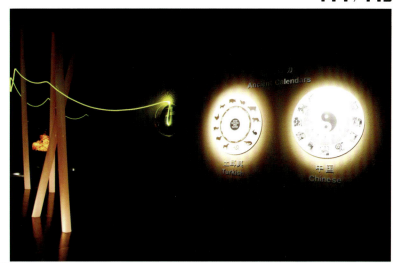

社团 · INSTITUTION

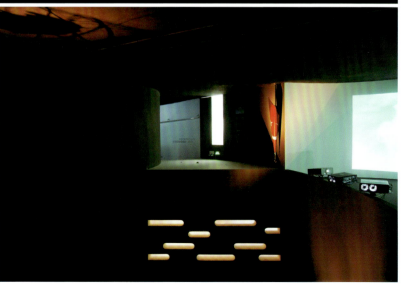

社团 · INSTITUTION

社团 · INSTITUTION

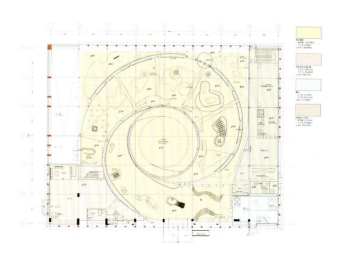

社团 · INSTITUTION

JERSEY BOYS THEATRE
JERSEY BOYS THEATRE

项目资料:
设计单位: HKS Inc.
摄影师: Blake Marvin, HKS, Inc.

Project Information:
Design Unit: HKS Inc.
Photographer: Blake Marvin, HKS, Inc.

社团 · INSTITUTION

HJØRRING CENTRAL LIBRARY

HJØRRING CENTRAL LIBRARY

项目资料：
设计单位：Bosch & Fjord
设计师：Bosch & Fjord
摄影师：Laura Stamer
客户：Hjørring Central Library
项目地址：stergade 30
面积：4900m²
主要材料：漆布, MDF, 环氧树脂地板、织物、泡沫、各式家具

Project Information：
Design Unit: Bosch & Fjord
Designer: Bosch & Fjord
Photographer: Laura Stamer
Client: Hjørring Central Library
Project Address: stergade 30
Area: 4900sqm
Materials: Linoleum, MDF, epoxy floor, textile, foam, various pieces of furniture

社团 · INSTITUTION

上海世博会ALG-ERIAN国家馆

NATIONAL ALGERIAN PAVILION AT WORLD EXPO SHANGHAI

项目资料：
设计单位：Totems Communication
创意总监：Peter van Lier
建筑团队：Melle Pama, Mark Niewenhuis
运动设计：Ammograph and Sugar Films
摄影师：Manfred H. Vogel
项目地址：中国上海

Project Information：
Design Unit: Totems Communication
Creative Director: Peter van Lier
Architect Team: Melle Pama, Mark Niewenhuis
Motion design: Ammograph and Sugar Films
Photographer: Manfred H. Vogel
Project Address: Shanghai, China

APPLEMORE COLLEGE PR CHOICE

APPLEMORE COLLEGE PR CHOICE

项目资料：
设计单位：Ippolito Fleitz Architects Gmbh

Project Information:
Design Unit: Ippolito Fleitz Architects Gmbh

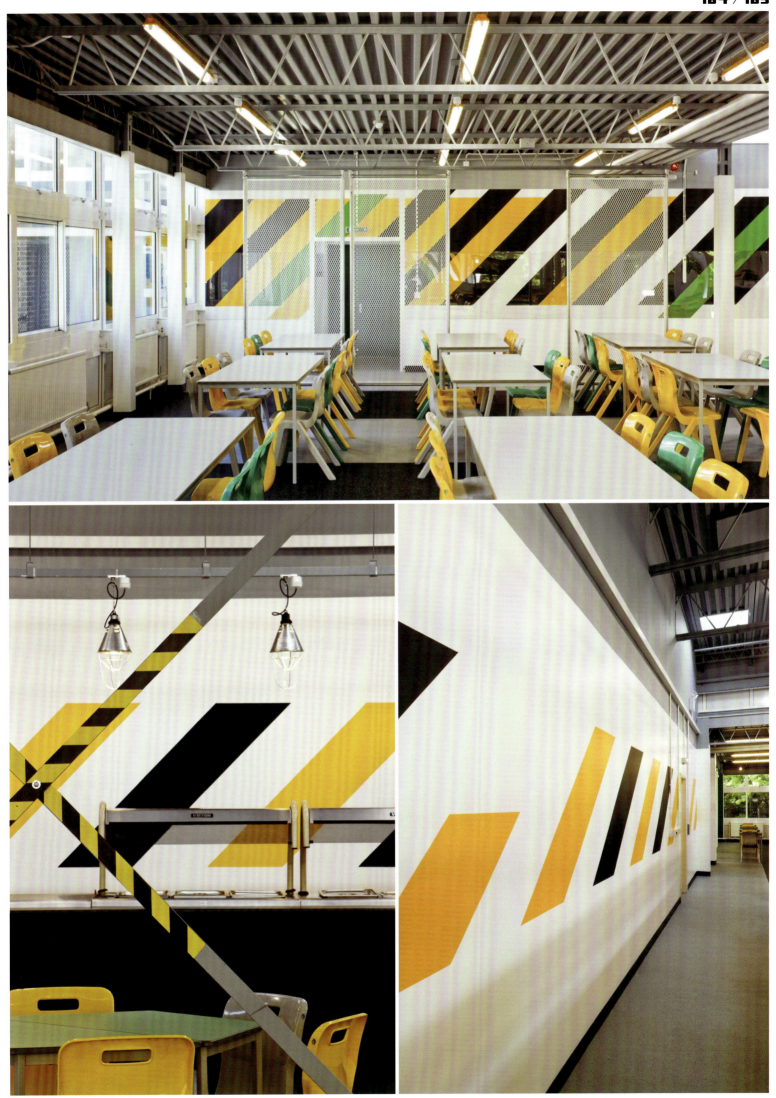

社团 · INSTITUTION

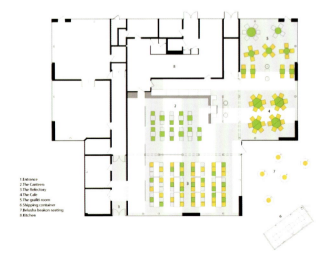

1. Entrance
2. The Canteen
3. The Refectory
4. The Cafe
5. The graffiti room
6. Shipping container
7. Belusha beakon seating
8. Kitchen

DORNIER MUSEUM
DORNIER MUSEUM

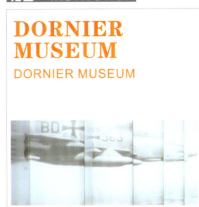

项目资料：
设计单位：ATELIER BRÜCKNER
设计师：Uwe R. Brueckner
摄影师：Johannes Seyerlein

Project Information:
Design Unit: ATELIER BRÜCKNER
Designer: Uwe R. Brueckner
Photographer: Johannes Seyerlein

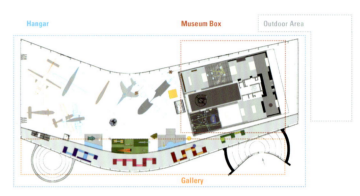

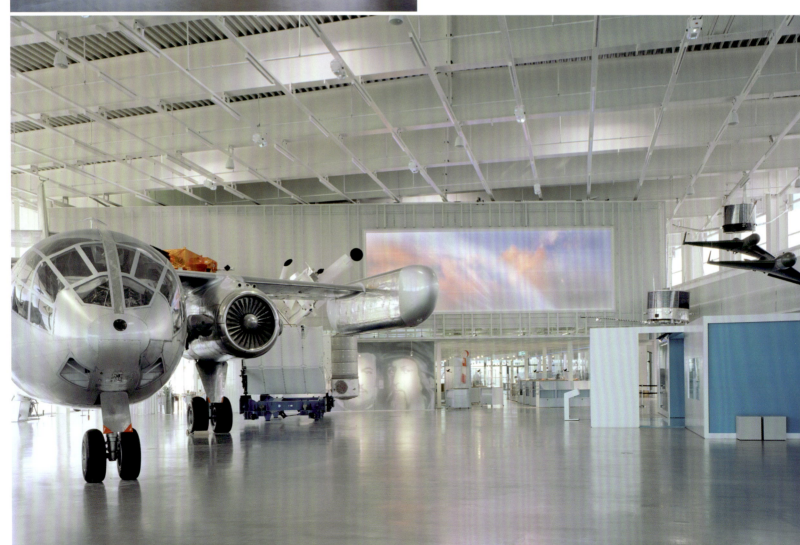

社团 · INSTITUTION

OBUSE LIBRARY MACHITOSHO TERRASOW

OBUSE LIBRARY
MACHITOSHO TERRASOW

项目资料:
设计单位: Japan NASCA Ltd.
设计师: Nobuaki Furuya
建筑团队: NASCA
结构设计: Masato ARAYA, Eijiro MORI, Saburo TAKAMA
摄影师: Satoshi Asakawa
项目地址: Obuse, Nagano
面积: 10511.44m²

Project Information:
Design Unit: Japan NASCA Ltd.
Designer: Nobuaki Furuya
Architect Team: NASCA
Structure Design: Masato ARAYA, Eijiro MORI, Saburo TAKAMA
Photographer: Satoshi Asakawa
Project Address: Obuse, Nagano
Area: 10511.44sqm

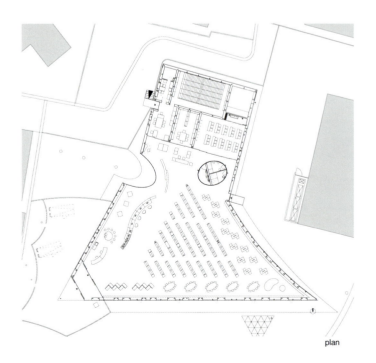

plan

社团 · INSTITUTION

ROC CARE
ROC CARE

项目资料：
设计单位：Tjep.

Project Information:
Design Unit: Tjep.

AN ABANDONED MOVIE PROP
AN ABANDONED MOVIE PROP

项目资料：
设计单位：Studio Ramin Visch
项目团队：Ramin Visch, Mark Helder (constructie), Peter van de Geer (akoestiek), Tobias Cassel, Femke Poppinga, Rick Abbenbroek
摄影师：Jeroen Musch & Rene Mesman
客户：Foundation Het Ketelhuis
面积：645m²

Project Information:
Design Unit: Studio Ramin Visch
Project team: Ramin Visch, Mark Helder (constructie), Peter van de Geer (akoestiek), Tobias Cassel, Femke Poppinga, Rick Abbenbroek
Phtographer: Jeroen Musch & Rene Mesman
Client: Foundation Het Ketelhuis
Area: 645sqm

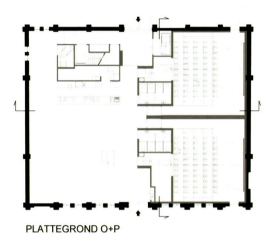

PLATTEGROND O+P

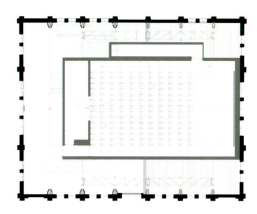

EERSTE VERDIEPING 5000+P

社团 · INSTITUTION

在第一次世界大战中的欧洲华工
CHINESE LABOURERS IN WORLD WAR I, EUROPE

项目资料：
设计单位：Totems Communication
创意总监：Peter van Lier
建筑团队：Simone Königshausen, Melle Pama
平面设计：Go Gumtree
摄影师：Hans Roggen
项目地址：比利时伊普雷斯

Project Information:
Design Unit: Totems Communication
Creative Director: Peter van Lier
Architect Team: Simone Königshausen, Melle Pama
Graphic design: Go Gumtree
Photographer: Hans Roggen
Project Address: Ypres, Belgium

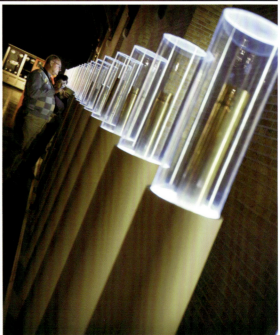

社团·INSTITUTION

THE WALT DISNEY FAMILY MUSEUM

THE WALT DISNEY FAMILY MUSEUM

项目资料：
设计单位：Rockwell Group
摄影师：Cesar Rubio

Project Information:
Design Unit: Rockwell Group
Photographer: Cesar Rubio

社团 · INSTITUTION

KUA TRY-OUT LAB

KUA TRY-OUT LAB

项目资料：
设计单位：Bosch & Fjord
设计师：Bosch & Fjord
摄影师：Anders Sune Berg
客户：哥本哈根大学
项目地址：Njalsgade 80
主要材料：壁毯、家具 (金属、木材、层压板)、地板 (聚氨酯)、照明

Project Information：
Design Unit: Bosch & Fjord
Designer: Bosch & Fjord
Photographer: Anders Sune Berg
Client: University of Copenhagen
Project Address: Njalsgade 80
Materials: Tapestry, furniture (metal, wood, laminate), floor (polyurethan), lighting

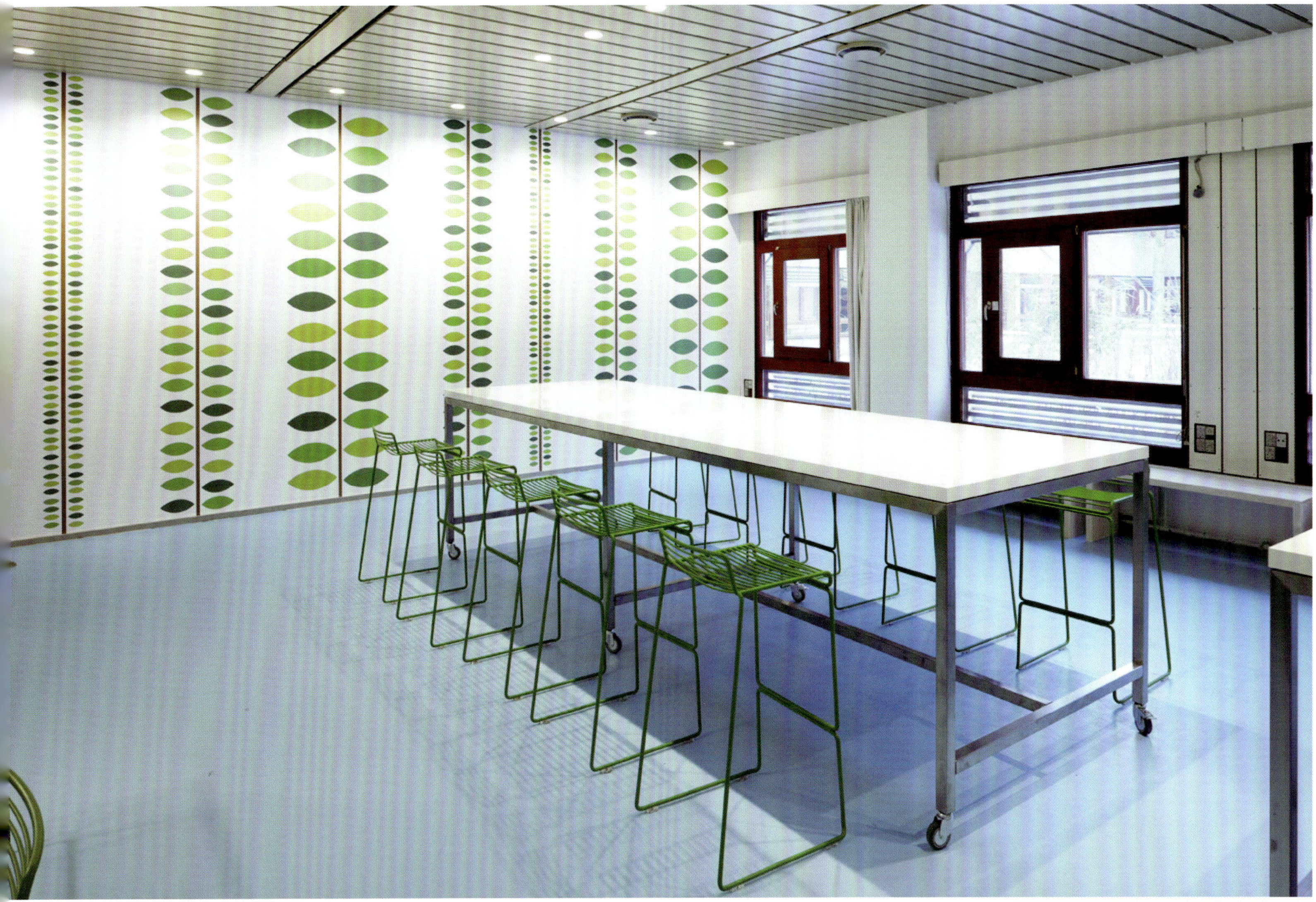

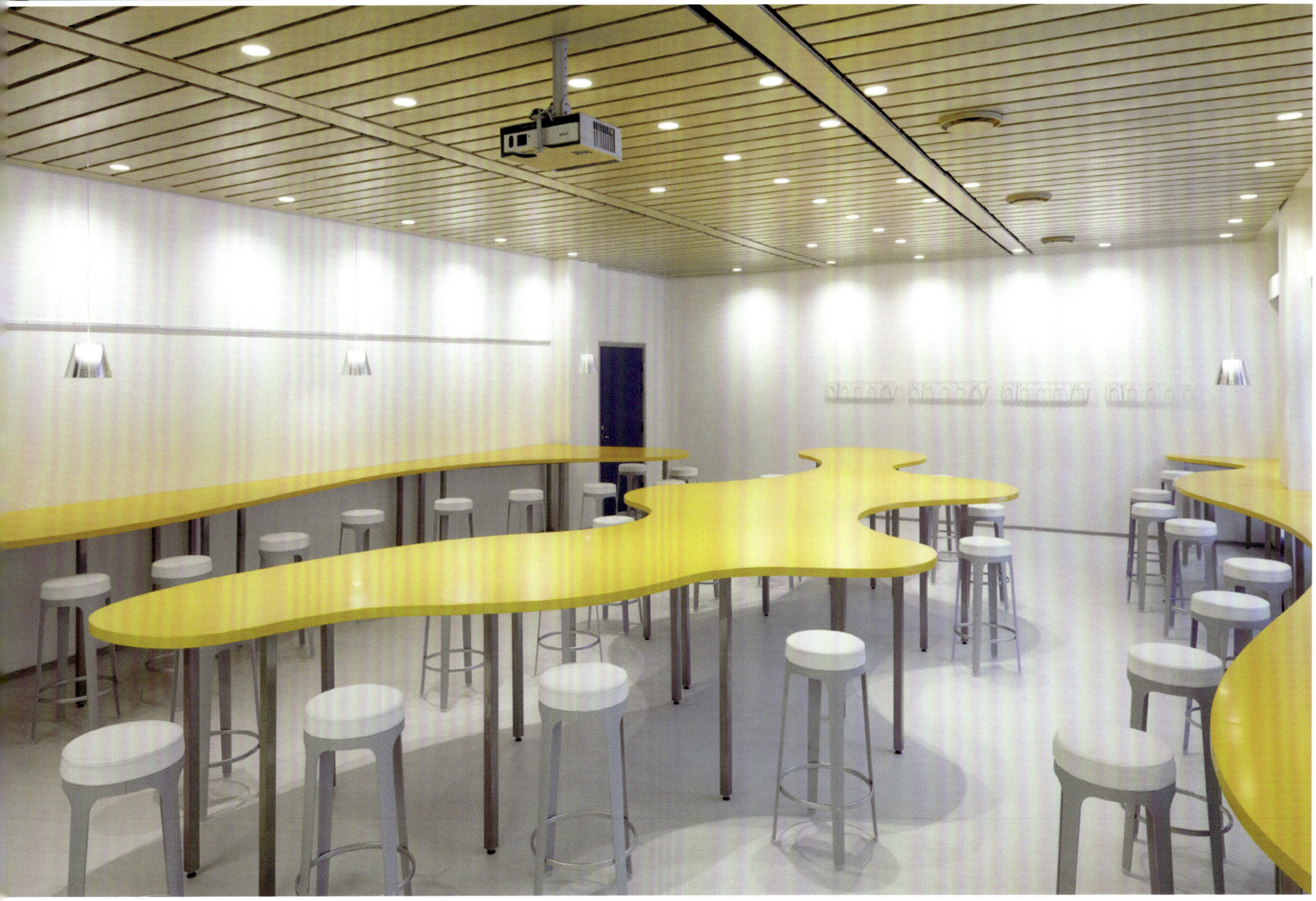

TOKYO BABY CAFÉ

TOKYO BABY CAFÉ

项目资料：
设计单位：nendo

Project Information:
Design Unit: nendo

社团 · INSTITUTION

音乐和戏剧学院
CONSERVATOIRE DE MUSIQUE ET D'ART DRAMATIQUE

项目资料：
设计单位：SAIA BARBARESE TOPOUZANOV ARCHITECTS
主设计师：MARIO SAIA
项目总监：DINO BARBARESE
客户：魁北克大学

Project Information：
Design Unit: SAIA BARBARESE TOPOUZANOV ARCHITECTS
Main Designer: MARIO SAIA
Project Director: DINO BARBARESE
Client: UNIVERSITY OF QUEBEC

社团 · INSTITUTION

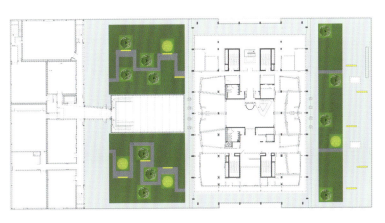

社团 · INSTITUTION

嘉禾青衣电影城
GOLDEN HARVEST TSING YI MULTIPLEX

项目资料：
设计单位：DPWT 设计公司
设计团队：陈轩明 吴永利 陈斌 伍蔼贤
摄影师：陈志威
客户：嘉禾娱乐事业有限公司
项目地址：香港新界青衣青衣城地下
面积：400m²
设计风格：抽象与极简主义
主要材料：喷漆玻璃、半透明薄膜、黑色不锈钢
完成时间：2009年03月

Project Information:
Design Unit: DPWT Design Ltd.
Participant Designer: Arthur Chan, Willie Wu, Bevin Chen and Alice Ng
Photographer- Mr. Diamond Chan
Client: Golden Harvest Entertainment Company Limited
Project Address: G/F, Maritime Square, Tsing Yi, NT, Hong Kong
Area: 400sqm
Design Style: Abstract and Minimalism
Materials: Spray Paint Glass, Film Membrane, and Black Stainless Steel
Completion: March 2009

社团 · INSTITUTION

JAQUA

项目资料：
设计单位：ZGF Architects LLP

Project Information:
Design Unit: ZGF Architects LLP

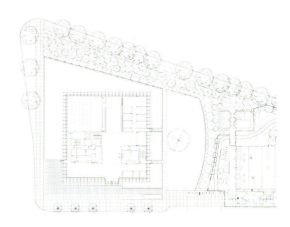

社团 · INSTITUTION

DESIGN LAB
DESIGN LAB

项目资料：
设计单位：Bosch & Fjord
设计师：Bosch & Fjord
摄影师：Anders Sune Berg
客户：IT University of Copenhagen
项目地址：Rued Langgaards Vej 7
面积：151m²
主要材料：中密度纤维板、平台建筑、层压板、油漆、家具

Project Information:
Design Unit: Bosch & Fjord
Designer: Bosch & Fjord
Photographer: Anders Sune Berg
Client: IT University of Copenhagen
Project Address: Rued Langgaards Vej 7
Area: 151sqm
Materials: Mdf platform construction, laminate, paint and furniture

韦斯特切斯特教室
WESTCHESTER REFORM TEMPLE & SCHOOL

项目资料：
设计单位：ROGERS MARVEL ARCHITECTS, PLLC
建筑团队：Rogers Marvel
摄影师：©Paul Warchol
项目地址：纽约斯卡斯代尔

Project Information：
Design Unit: ROGERS MARVEL ARCHITECTS, PLLC
Architecture Team: Rogers Marvel
Photographer: ©Paul Warchol
Project Address: Scarsdale, NY

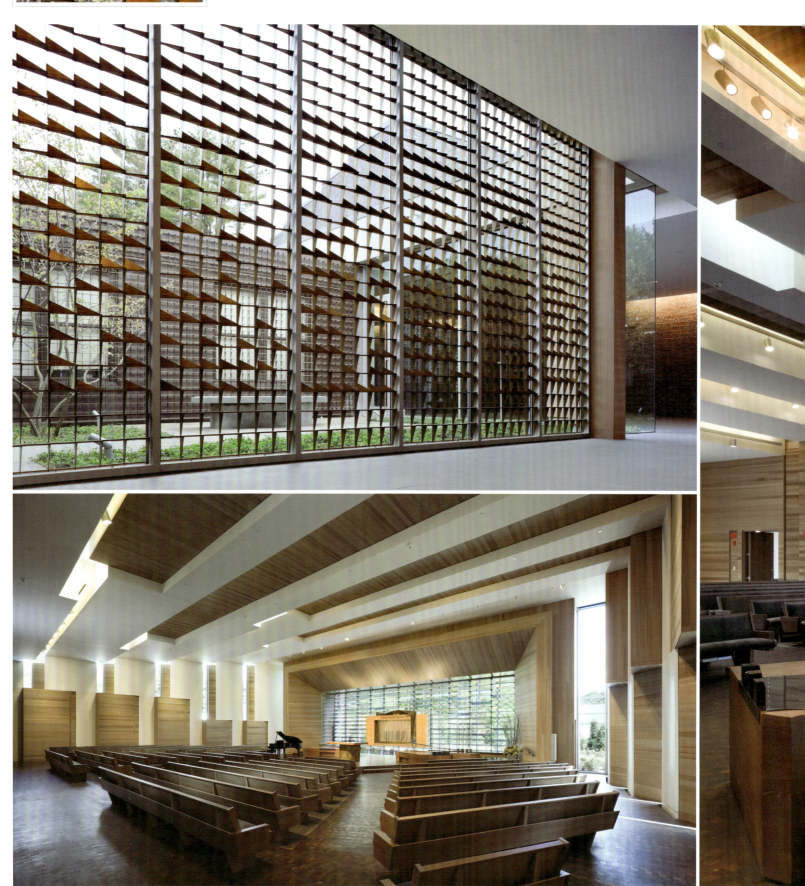

嘉禾荃新天地电影城
GH CITYWALK CINEMA

项目资料：
设计单位：DPWT 设计公司
设计团队：陈轩明 吴永利 陈斌 伍蔼贤
摄影师：陈志威
客户：橙天嘉禾娱乐集团
项目地址：香港新界荃湾荃新天地第二期1楼
面积：1440m²
设计风格：现代、有活力、启发灵感
主要材料：白色玻璃墙、灰镜、碳色乙烯基地板、黑色地毯、不锈钢、重点强调色彩鲜明的隔音壁板与转色发光二极管照明
完成时间：2009年12月

Project Information:
Design Unit: DPWT Design Ltd.
Designer: Arthur Chan, Willie Wu, Bevin Chen and Alice Ng
Photographer: Mr. Diamond Chan
Client: Orange Sky Golden Harvest Entertainment Group
Address: 1/F, Citywalk 2, Tsuen Wan, New Territories, Hong Kong
Area: 1440sqm
Design Style: Modern, Energetic, and Inspiring
Materials: White tempered glass, grey mirror, charcoal vinyl flooring, black carpet, stainless steel and vibrant colours' fabric acoustic wall panel with changing colours LED lighting for accent highlight.
Completion: December 2009

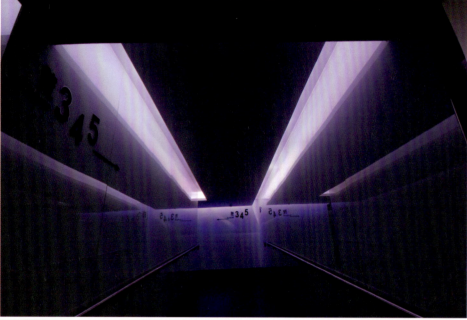

COWBOYS STADIUM

COWBOYS STADIUM

项目资料:
设计单位: HKS Inc.
摄影师: Blake Marvin, HKS, Inc.

Project Information:
Design Unit: HKS Inc.
Photographer: Blake Marvin, HKS, Inc.

社团·INSTITUTION

龙的DNA
DNA OF DRAGON

项目资料:
设计单位: 天坊室内计划有限公司
设计总监（主创）: 张清平
参与设计团队: 王美智 陈可津 吴怡芝 柯玫绮
摄影师: 刘俊杰
项目地址: 台湾台中

Project Information:
Design Unit: Tianfang Interior Planning Co., Ltd.
Design Director(Main Director): Zhang Qingping
Involed Design Team: Wang Meizhi, Chen Kejin, Wu Yizhi, Ke Meiqi
Photographer: Liu Junjie
Project Address: Taiwan Taizhong

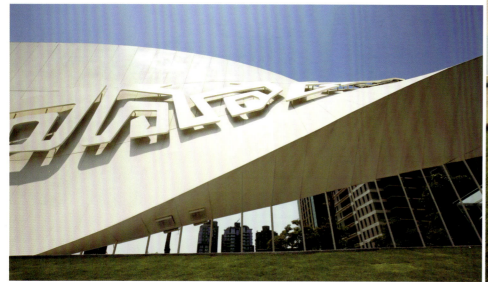

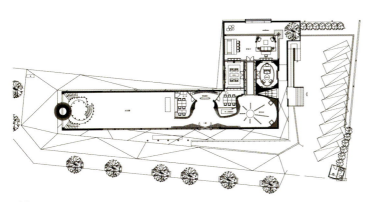

社团 · INSTITUTION

社团 · INSTITUTION

社团 · INSTITUTION